Written by Molly Sanchez and
The Good and the Beautiful Team
Designed by Anna Asfour

© 2022 Jenny Phillips | goodandbeautiful.com

In time past, the world was filled with awe-inspiring creatures large and small. While many of these animals can no longer be found, they left behind clues about their existence on the earth. From teeth to toes, scientists study these pieces like a puzzle to give us a glimpse into the past.

New discoveries are constantly found, and new speculations are made about these ancient animals as we gather more information. The animals featured in this book are rendered from a compilation of the best information scientists have available today.

# DIMETRODON

Dimetrodon, often mistaken for a dinosaur, was a mammal-like creature that fed on fish, reptiles, and amphibians. Since its initial discovery in 1878, twenty species have been identified. This animal walked on four legs and had a magnificent sail down its back. Many scientists believe that Dimetrodon's sail was used to help regulate its temperature, though other studies indicate that this may not have been the case.

## FUN FACT:

Dimetrodon means "two measures of teeth." It had long teeth in the front and shorter teeth in the back.

NORTH AMERICA

## INFO:

**SIZE:**

1.83–4.57 m (6–15 ft) long

**WEIGHT:**

27.22–249.48 kg (60–550 lb)

**CARNIVORE**

**FOUND:**

North America

## INFO:

**SIZE:**
99.06 cm (39 in) tall

**WEIGHT:**
158.76–281.23 kg (350–620 lb)

**CARNIVORE**

**FOUND:**
North and South America

## FUN FACT:

The Smilodon's mouth could open up 120 degrees to make room for its huge teeth to bite. Modern lions can open their jaws 65 degrees.

NORTH AMERICA

SOUTH AMERICA

# SMILODON

Also known as a saber-toothed tiger, this fierce animal preyed on animals such as bison and camels. Based upon studies of Smilodon's body structure and size, this muscular animal is believed to have hidden in forests and bushes while hunting its next meal. The most prominent feature of the Smilodon was its canine teeth, which were more than 25.4 cm (10 in) long from root to tip.

# MEGALONYX JEFFERSONII

Megalonyx jeffersonii, an ancient ground sloth, lived mostly in hardwood forests and woodlands. This heavily built animal had a massive jaw, short snout, and large teeth. Its hind limbs were flat-footed, allowing it to stand up and reach for leaves with its powerful front claws.

## FUN FACT:

An even larger, elephant-sized sloth existed in South America called Megatherium.

NORTH AMERICA

## INFO:

**SIZE:**
3 m (9.8 ft) long

**WEIGHT:**
Up to 998.9 kg (2,200 lb)

**HERBIVORE**

**FOUND:**
North America

## INFO:

**SIZE:**

3 m (9.8 ft) long

1.5 m (5 ft) high

**WEIGHT:**

907.18–1,814.37 kg
(2,000–4,000 lb)

**HERBIVORE**

**FOUND:**

South America

## FUN FACT:

A Josephoartigasia monesi's 53.34 cm (21 in) long skull was found embedded in a boulder on a Uruguayan beach.

SOUTH AMERICA

# JOSEPHOARTIGASIA MONESI

Josephoartigasia monesi was a massive rodent, speculated to have feasted on aquatic plants and fruits based on the size and shape of its teeth. Its large incisors were used for defense against predators, as well as for rooting around on the ground for food.

12-inch Incisor

ANCIENT ANIMALS

# TITANOBOA

Weaving its massive body through hot, humid jungles, the Titanoboa was an enormous hunter that could swallow its prey whole. Based on scientific research, the snake resembled a boa constrictor yet hunted like a crocodile, ambushing its prey instead of wrapping around and squeezing its meal.

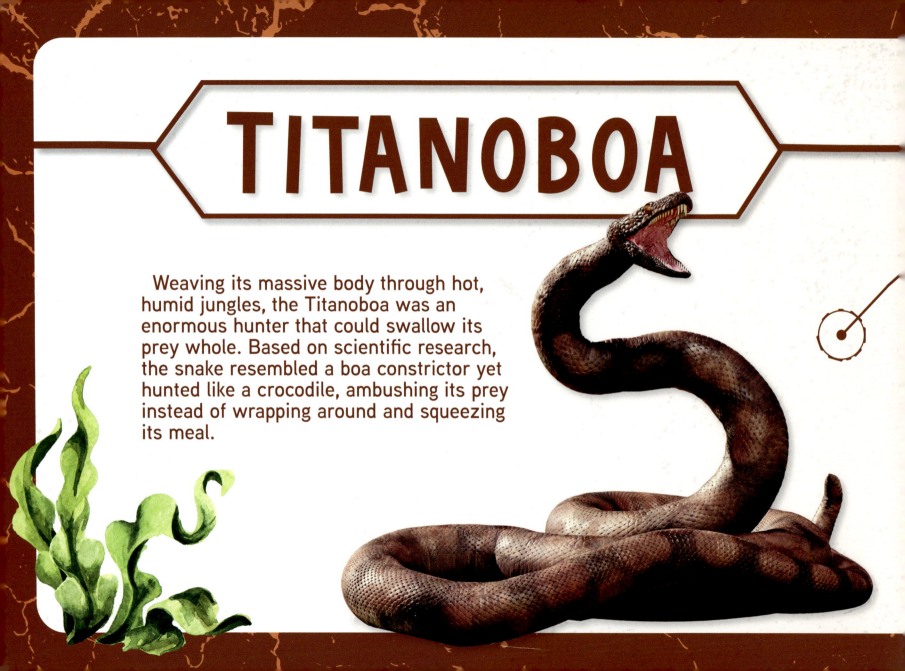

## FUN FACT:

Titanoboa could go nearly a year between meals. Its body was 91.44 cm (3 ft) in diameter.

COLOMBIA

## INFO:

**SIZE:**
15.24 m (50 ft) long

**WEIGHT:**
About 1,133.98 kg (2,500 lb)

**CARNIVORE**

**FOUND:**
Colombia

ANCIENT ANIMALS

## INFO:

**SIZE:**
1.71 m (5.6 ft)

**WEIGHT:**
Unknown

**CARNIVORE**

**FOUND:**
Crater in Iowa, USA

## FUN FACT:

Pentecopterus decorahensis is named after the penteconter, an ancient Greek warship.

CRATER IN IOWA

# PENTECOPTERUS DECORAHENSIS
## (GIANT SWIMMING SCORPIONS)

This underwater arthropod is actually an ancient relative of today's spiders, ticks, and lobsters. The very well-preserved fossil of Pentecopterus was found in Iowa, USA, in 2010. The Pentecopterus, or "sea scorpion," ranged from quite small to over 1.5 m (6 ft) long and had tiny hairs on its legs. With large grasping limbs, Pentecopterus was a fierce aquatic predator in its time.

# DUNKLEOSTEUS
## (GIANT FISH)

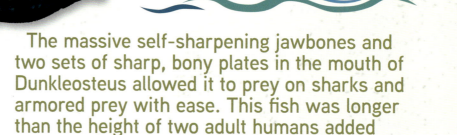

The massive self-sharpening jawbones and two sets of sharp, bony plates in the mouth of Dunkleosteus allowed it to prey on sharks and armored prey with ease. This fish was longer than the height of two adult humans added together, and its head was covered in thick armor. It was probably a slow but powerful swimmer.

## FUN FACT:

Dunkleosteus had a bite force of 611.9 kg (1,349 lb), thanks to a unique four-part jaw mechanism.

## INFO:

**SIZE:**

6 m (20 ft)

**WEIGHT:**

900+ kg (2,000+ lb)

**CARNIVORE**

**FOUND:**

North America and Europe

## INFO:

**SIZE:**
Up to 11.89 m (39 ft)

**WEIGHT:**
Up to 4,989.52 kg (11,000 lb)

**CARNIVORE**

**FOUND:**
Mexico, 10 states in the USA; most abundant in Georgia

## FUN FACT:

Deinosuchus possibly lived to be 50 years old. It was large enough to have preyed on dinosaurs.

USA

MEXICO

# DEINOSUCHUS
## (MEGA CROCODILE)

A close relative of the alligator, Deinosuchus was larger than most other predators in the subtropical seawater that used to cover the state of Ohio, USA. A cunning predator, Deinosuchus would wait at the shoreline of the sea, ambush its prey, and drag it into the water to submerge it before eating it.

# ANOMALOCARIS
## (GIANT SHRIMP)

In Greek, anomalocaris means "unusual shrimp." At a length of around 91.44 m (3 ft) long, this ancient arthropod was certainly unusual. Anomalocaris was able to swim at great speeds due to its undulating motion, and it had sharp spikes on its limbs, useful for grabbing prey. With 32 overlapping plates in its mouth, Anomalocaris may have been able to crush the thick armor on food like trilobites, but it has also been theorized that it was a filter feeder, using its spiky arms to sift through gravel for smaller prey.

## FUN FACT:

Anomalocaris' large stalked eyes had 16,000 lenses each, which gave it 360-degree eyesight.

## INFO:

**SIZE:**
Up to 91.44 cm (3 ft)

**WEIGHT:**
9.07 kg (20 lb)

**CARNIVORE**

**FOUND:**
Utah USA, Canada, China, Australia, and Greenland

## INFO:

**SIZE:**
3.35–3.66 m (11–12 ft) tall standing;
1.52–1.83 m (5–6 ft) long (head to tail)

**WEIGHT:**
900+ kg (2,000+ lb)

**OMNIVORE**

**FOUND:**
North America

## FUN FACT:

Arctodus could run 48–64 kph (30–40 mph), but not for very long. Males of this species were larger than females.

NORTH AMERICA

# ARCTODUS
## (GIANT SHORT-FACED BEAR)

Arctodus was much larger than today's grizzly or polar bears with long legs and a relatively short body and snout. Possibly an omnivore, Arctodus probably ate a wide range of meats and vegetation, but some scientists have argued that it may have been a scavenger. Although its long legs would suggest that it was a fast runner, its bulk kept it from being able to turn or stop quickly. Arctodus was most likely better at endurance running at a steady pace over long distances.

# MEGALODON

Megalodon was the largest shark and one of the biggest fish that has ever lived. It's no wonder the word megalodon means "giant teeth" in Greek, as one of the longest Megalodon teeth ever found was a whopping 16.5 cm (6.5 in). Megalodon was at the top of the ocean food chain, chowing on dolphins, fish, seals, whales, sharks, and other marine animals with its 276 teeth. If it were alive today, Megalodon would be much bigger than a great white shark—by about 12 m (39 ft) in length and 25,000 kg (55,100 lb) in weight.

## FUN FACT:

Females were larger than males. Megalodons had jaw openings of 3 m (9.8 ft), top to bottom. Teeth were nearly 17.78 cm (7 in) long and were much larger than shark teeth.

- NORTH AMERICA
- EUROPE
- ASIA
- AFRICA
- SOUTH AMERICA
- AUSTRALIA

## INFO:

**SIZE:**
10–25 m (33–82 ft)

**WEIGHT:**
29,937.1–64,863.71 kg (66,000–143,000 lb)

**CARNIVORE**

**FOUND:**
Coastlines of every continent except Antarctica

## INFO:

**SIZE:**
6–7 m (19.6–23 ft) wingspan

**WEIGHT:**
21.77–39.92 kg (48–88 lb)

**CARNIVORE**

**FOUND:**
South Carolina, USA

## FUN FACT:

This ancient bird's "teeth" were sharp extensions of bone from the jaw.

SOUTH CAROLINA

# PELAGORNIS SANDERSI

With a wingspan of at least 6–7 m (19.6–23 ft), Pelagornis sandersi was likely able to soar over the ocean for hours with barely a flap of its wings. It likely ate fish and squid from the surface of the water, piercing its prey with toothlike spiky structures in its mouth. Although Pelagornis had such an impressive wingspan, it also had stumpy, short legs and most likely could only take off by jumping off cliffs or running downhill.

# MEGALANIA
## (VARANUS PRISCUS)

Categorized in the monitor family of lizards along with today's Komodo dragon, fossils from the gigantic Megalania lizard have been found all over Australia. At about 5 m (16.4 ft) long and with sharp, curved teeth, Megalania would have been able to take down large prey such as pygmy elephants (now extinct), kangaroos, and tortoises. It is believed that Megalania also carried venom in its bite.

## FUN FACT:

Megalania is the largest known land lizard. The Komodo dragon is one of its closest relatives.

AUSTRALIA

## INFO:

**SIZE:**
3.5–7 m (11.5–23 ft)

**WEIGHT:**
Up to 1,940.02 kg (4,277 lb)

**CARNIVORE**

**FOUND:**
Australia

## INFO:

**SIZE:**
Up to 3 m (9.8 ft)

**WEIGHT:**
Unknown

**CARNIVORE**

**FOUND:**
South America

## FUN FACT:

The Phorusrhacid were fast runners, up to 48 kph (30 mph). They likely chased down rodents for prey.

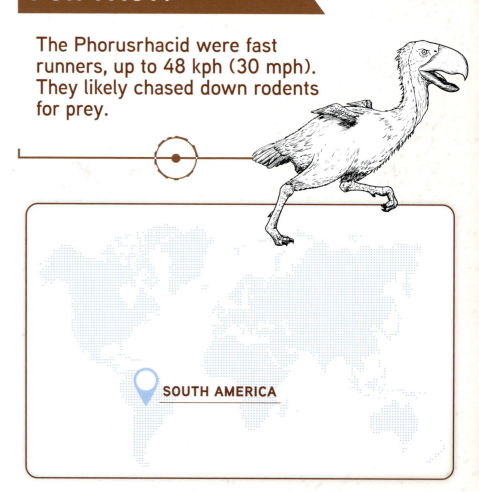

SOUTH AMERICA

# PHORUSRHACID

Nicknamed the "terror bird" for its predatory feeding habits and ferocious hooked beak, the Phorusrhacid was an apex hunter that lived on the continent of South America. Standing between 1–3 m (3.3–9.8 ft) tall, scientists speculate that this long-legged bird used its height to intimidate prey and its large beak to strike downward at the animals. It may also have held its prey down with sharp talons and pecked at the prey with its powerful beak.

# MEGACEROPS

Greek for "giant horned face," the name Megacerops perfectly fits the lumbering beast with two blunt horns on its snout. Megacerops was a herbivore, probably feeding on soft vegetation and fruit. When it came to finding a mate, a male Megacerops would use its horns to challenge and fight other males for dominance. They are also known as Brontotherium, which means "thunder horse" in Greek and describes their close genetic relation to modern day horses.

## FUN FACT:

This ancient animal resembled a rhinoceros but is more closely related to a horse. Its front feet had four toes, and its hind feet had three.

NORTH AMERICA

## INFO:

**SIZE:**
2.5 m (8.2 ft) tall
5 m (16 ft) long

**WEIGHT:**
3,265.87 kg (7,200 lb)

**HERBIVORE**

**FOUND:**
North America

## INFO:

### SIZE:
4.88 m (16 ft) tall at its shoulder

7.92 m (26 ft) long

### WEIGHT:
19,958+ kg (44,000+ lb)

### HERBIVORE

### FOUND:
Eurasia

## FUN FACT:

When discovered in China by farmers, they thought they had found dragon bones. Some of the bones were crushed and used in Chinese medicine.

EURASIA

# PARACERATHERIUM LINXIAENSE

It might be hard to imagine an animal much bigger than an elephant, but Paraceratherium Linxiaense was actually the size of five elephants. In fact, this giant herbivore is the largest land mammal archaeologists have discovered as of today. Paraceratherium linxiaense had a skull shape similar to that of modern rhinos, which was well-suited for grabbing vegetation from tall trees, and long legs which helped it traverse long distances easily.

# ANDREWSARCHUS

By studying the skull of the Andrewsarchus, scientists have theorized that this animal was either a hunter or a scavenger. It appears to have had characteristics such as a very strong jaw and wide cheekbones. Not much else is known about Andrewsarchus, but perhaps one day more fossils from this meat-eating creature will be found.

**FUN FACT:** The only fossil of this species ever uncovered is a .8-m (2.6-ft) long skull found in Mongolia. From this specimen, scientists have only been able to make educated guesses about what the rest of the body may have looked like.

## INFO:

### SIZE:
About 1.77 m (5.8 ft) tall at the shoulder

3.35 m (11 ft) long

### WEIGHT:
Maybe 907.18–1,814.37 kg (2,000–4,000 lb)

### CARNIVORE OR OMNIVORE

### FOUND:
Mongolia

ANCIENT ANIMALS

## INFO:

**SIZE:**
About 1.5 m (5 ft) tall
3.35 m (11 ft) long

**WEIGHT:**
Up to 1,995.81 kg (4,400 lb)

**HERBIVORE**

**FOUND:**
South America

## FUN FACT:

Glyptodon was similar in size and weight to a Volkswagen Beetle. It used its bone-studded tail to fight its own species for territory and/or mating rights.

NORTH AMERICA

SOUTH AMERICA

# GLYPTODON

Covered from head to toe in a thick layer of bony armor, Glyptodon might remind one of an animal living on Earth today. Armadillos are smaller relatives of this giant extinct mammal, which lived in the swamps of North and South America. Glyptodon's protective armor would have made it very hard for predators to attack and eat it.

# PALAEEUDYPTES
## (GIANT PENGUINS)

Including four species of large ancient penguins, fossils of the Palaeeudyptes genus of birds have been found on Seymour Island, Antarctica and in New Zealand. The largest of the four species was taller than most humans at about 2 m (6.7 ft) tall, while the smallest was closer to the size of modern emperor penguins. Based on their size, scientists think these "colossus penguins" were able to stay underwater for around 40 minutes at a time.

## FUN FACT:

A similar penguin today is the emperor penguin at 1.22 m (4 ft) tall and close to 45 kg (100 lb).

NEW ZEALAND

ANTARCTICA

## INFO:

**SIZE:**

1.83 m (6 ft) tall

**WEIGHT:**

About 113.4 kg (250 lb)

**CARNIVORE**

**FOUND:**

Antarctica

ANCIENT ANIMALS

## INFO:

**SIZE:**
Up to 91.44 cm (3 ft) long

**WEIGHT:**
9.98 kg (22 lb)

**CARNIVORE OR HERRBIVORE**

**FOUND:**
Argentina

## FUN FACT:

Scientists estimate that Megapiranha was at least ten times as big as modern piranhas.

ARGENTINA

# MEGAPIRANHA

Only the teeth from this ancient fish have been found, making it difficult to decipher what this fish looked like and what it ate. What scientists do know is that the teeth of the Megapiranha were arranged in a zig-zag pattern across the front jaw and could have been useful for either a carnivorous or herbivorous fish species.

ANCIENT ANIMALS

# Hieraaetus moorei
## (Haast's eagle)

This magnificent creature is the largest eagle ever known to exist, with the females being much larger than the males. The Hieraaetus moorei lived on the island of New Zealand, where their sharp talons and curved beaks enabled them to feast on flightless birds and eggs. Scientists speculate that their muscular legs and wing muscles allowed them to take off in flight from the ground.

**FUN FACT:** Haast's eagle went extinct only about 600 years ago. Scientists theorize it went extinct because people began hunting the moa, which was its prey.

NEW ZEALAND

## INFO:

**SIZE:**
About 91.44 cm (3 ft) tall
Up to 3 m (9.8 ft) wingspan

**WEIGHT:**
Up to 14.97 kg (33 lb)

**CARNIVORE**

**FOUND:**
New Zealand

ANCIENT ANIMALS

## INFO:

**SIZE:**
1.74 m (5.7 ft) at the shoulder

**WEIGHT:**
Unknown

**OMNIVORE**

**FOUND:**
North America

## FUN FACT:

Daeodon shoshonensis are related to pigs and hippos.

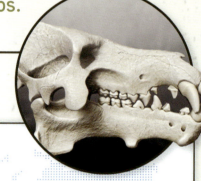

NORTH AMERICA

# DAEODON SHOSHONENSIS

With its name originating from the Greek words "dreadful teeth," this intimidating creature feasted on a variety of nuts, roots, and plants, as well as scavenged meat and bones. Daeodon's skull was more than .91 m (3 ft) long, and it walked on two-toed feet; therefore scientists believe this animal was bulky and awkward as it roamed the land.

# THALASSOCNUS

Thalassocnus, a genus of semiaquatic ground sloths, relied on their large paws for paddling more so than aggressive swimming. With long snouts and internal nostrils far into its head, this animal was able to sink under water and walk across the seafloor to dig up its next meal of seaweed and seagrass. Similar to a beaver's tail, Thalassocnus's long tail was most likely used to aid in diving and balance.

## FUN FACT:

Thalassocnus's dense bones helped it to sink so it could walk the ocean floor in search of food.

PERU
CHILE

## INFO:

**SIZE:**
Up to 3.35 m (11 ft) long

**WEIGHT:**
Up to 3,628.74 kg (8,000 lb)

**HERBIVORE**

**FOUND:**
Peru and Chile

## INFO:

**SIZE:**
3.66 m (12 ft) tall with neck outstretched

**WEIGHT:**
231.33 kg (510 lb)

**HERBIVORE**

**FOUND:**
New Zealand

## FUN FACT:

The giant moa became extinct around AD 1400 after being overhunted.

NEW ZEALAND

# DINORNIS ROBUSTUS
## (GIANT MOA)

Standing at over 12 feet tall, this flightless bird may have been the tallest avian species to ever live. Feasting on twigs and leaves both low to the ground and high in the air, the giant moa could raise its long neck and use its strong beak to pluck the vegetation from tall trees. With a lifespan of 10-30 years, these magnificent creatures chose to live in small families, instead of a large flock like other birds.

# STUPENDEMYS

Stupendemys, the largest freshwater turtle known to exist, ate plants, animals, and nearly anything that would fit in its mouth. This side-necked turtle would fold its long neck into one side of its shell for protection, and its heavy weight allowed it to stay under water for extended periods of time. Scientists discovered that the male Stupendemys had horns on the upper part of its shell, which is called the carapace. These horns are speculated to have been used for defense.

**FUN FACT:**

Stupendemys translates to "stupendous turtle" which is an accurate description of this massive genus.

SOUTH AMERICA

**INFO:**

**SIZE:**
Up to 3.96 m (13 ft) long

**WEIGHT:**
1,133.98 kg (2,500 lb)

**OMNIVORE**

**FOUND:**
South America

ANCIENT ANIMALS

## INFO:

**SIZE:**
3 m (9.8 ft) long

**WEIGHT:**
1,043.26 kg (2,300 lb)

**HERBIVORE**

**FOUND:**
South America

## FUN FACT:

Macrauchenia may have had a trunk. Its leg bones indicate it could probably swerve or change direction abruptly to escape predators.

SOUTH AMERICA

# MACRAUCHENIA

The long-limbed, long-necked Macrauchenia had sturdy legs and a small head with a trunk-like protrusion called a proboscis. This creature's nostrils, however, are found up on its head, between its eyes. The location of its nostrils could have prevented them from getting irritated while it plucked leaves from high trees. This ancient animal was a herbivore, and its large size, muscular build, and quick-turning abilities made it difficult for predators to catch.

# GIGANTOPITHECUS

Gigantopithecus, a massive ancient ape, could be found lumbering through subtropical Asia. By studying teeth and jaw remains of this giant creature, scientists have concluded that their diets consisted of plants, leaves, and fruits. Tooth jaws that have been discovered show high rates of cavities and decay, most likely attributed to their high-sugar fruit diet.

## FUN FACT:

Many of Gigantopithecus's teeth have been recovered from pharmacies in China, where people thought they were dragon teeth with healing properties. The only remains found are teeth and parts of the jaw. Some like to connect Gigantopithecus to the folklorish Yeti or Bigfoot.

SOUTHEAST ASIA

## INFO:

**SIZE:**
Up to 3 m (9.8 ft) tall

**WEIGHT:**
Up to 498.95 kg (1,100 lb)

**HERBIVORE**

**FOUND:**
Southern China and southeast mainland Asia

ANCIENT ANIMALS

## INFO:

**SIZE:**
1 m (3.3 ft) long

**WEIGHT:**
4.54–15.88 kg (10–35 lb)

**CARNIVORE**

**FOUND:**
North America and Africa

## FUN FACT:

This amphibian likely fed on mollusks, crustaceans, and carrion. It burrowed into the ground to hibernate during times of drought.

# DIPLOCAULUS

With its long horns at the back of its skull creating a boomerang-shaped head, Diplocaulus was a distinctive creature. Some scientists have hypothesized Diplocaulus's horns were used as defensive features. Other scientists theorize that they were used to decrease drag and increase swimming speed by controlling how much water flowed over its head. Though this creature had four legs, its muscular tail is thought to have powered its movements through its swampy habitat.

# CHALICOTHERIUM

This odd-looking creature, Chalicotherium, had long clawed forelimbs and paw-like shorter hind limbs. Based on scientific studies, it is speculated that it walked on its front knuckles to protect its sharp claws from the ground. It would use its powerful forelimbs and claws to pull and scrape for food or to defend itself. Despite its fierce appearance, Chalicotherium's diet most likely consisted of leaves and other soft vegetation.

## FUN FACT:

Chalicotherium shares features with the great ape, panda, ground sloth, and the horse.

NORTH AMERICA

EURASIA

AFRICA

## INFO:

**SIZE:**
2.74 m (9 ft) high at the shoulder

**WEIGHT:**
Up to 907.18 kg (2,000 lb)

**HERBIVORE**

**FOUND:**
Africa, Eurasia, and North America

## INFO:

**SIZE:**
Up to 3.41 m (11.2 ft) tall at the shoulder

**WEIGHT:**
Up to 5,987.42 kg (13,200 lb)

**HERBIVORE**

**FOUND:**
Northern Eurasia and North America

## FUN FACT:

A baby woolly mammoth would weigh about 90.72 kg (200 lb) at birth. They lived for about 60 years. Woolly mammoths had small ears and short tails which helped prevent frostbite.

NORTHERN EURASIA

NORTH AMERICA

# WOOLLY MAMMOTH

The massive woolly mammoth is known for its long, thick, curved tusks that were used to break ice and scrape tree limbs, as well as to defend itself against predators. Its fat stores, short undercoat, and long fur kept it warm in the frigid lands it roamed. Scientists believe that this social animal likely lived in female-led family groups, similar to modern-day elephants. Living in a herd allowed for better protection against wolves and large felines.